DEDICATION

All of the pioneers, visionaries, and inventors influencing the direction of automation are honored in this book. To the engineers, researchers, and thought leaders whose tireless work in robotics is moving us into a new era of efficiency and possibilities. We are motivated to push the limits of technology by your commitment and enthusiasm.

Additionally, this work is devoted to those who welcome change with open minds, to companies and individuals that have the guts to adopt new technology and improve industries. I hope this book helps you on your path to a more automated and intelligent future.

DISCLAIMER

This book contains information that should only be used for general informative purposes. Although every attempt has been taken to guarantee the content's authenticity and dependability, the author and publisher make no guarantees or representations about the information's completeness, accuracy, or suitability for any particular purpose.

Technological developments, industry-specific factors, and unique business situations all influence the use of polyfunctional robotics and the tactics covered in this book. Before making decisions based on the information provided, readers are urged to speak with experts, professionals in the field, and legal counsel.

Any direct or indirect losses, damages, or repercussions resulting from the use or reliance on the material in this book are not the responsibility of the author or publisher. All company names, service marks, and trademarks belong to their respective owners.

Any references to goods or services in this book are purely

illustrative and do not advocate any particular brand or product.

CONTENTS

ACKNOWLEDGMENTS..1

CHAPTER 1..1

Overview of Multipurpose Robots..1

    1.1 Polyfunctional Robots: What Are They?.............................1

    1.2 Polyfunctional Robotics' Main Drivers................................4

    1.3 Comparing with Task-Specific Robots................................6

CHAPTER 2..10

Polyfunctional Robots' Technological Underpinnings...........10

    2.1 Essential Elements..10

    2.2 Machine Learning and AI's Role.......................................15

    2.3 Cloud computing and IoT integration................................18

CHAPTER 3..23

Industry-Wide Applications..23

    3.1 Production...23

    3.2 Medical Care..27

    3.3 Farming...31

CHAPTER 4..35

Polyfunctional Robot Design Principles...................................35

    4.1 Scalability and Modularity...35

    4.2 Human-Robot Interaction Ergonomics..............................39

    4.3 Sustainability and Energy Efficiency.................................42

CHAPTER 5..46

Polyfunctional Robotics Challenges..46

5.1 Technical Difficulties..46

5.2 ROI and Cost Factors..50

5.3 Concerns about Regulation and Ethics.......................................54

CHAPTER 6...58

Cooperation Between Humans and Robots................................. 58

6.1 Polyfunctional Robots' Function in Collaborative Work Environments...59

6.2 Safeguarding Human-Robot Communication...........................62

6.3 Adoption and Training..65

CHAPTER 7...70

Polyfunctional Robots' Economic Impact....................................70

7.1 Analysis of Cost-Benefit...70

7.2 Employment Impact..74

7.3 Strengthening SMEs (small and medium businesses)................... 77

CHAPTER 8...82

Multipurpose Robots in Intelligent Cities....................................82

8.1 Improving City Life..82

8.2 Logistics and Transportation...86

8.3 Green initiatives and sustainability..89

CHAPTER 9...93

Polyfunctional Robotics' Future Trends......................................93

9.1 New Technologies Fueling Creativity....................................... 94

9.2 Polyfunctional Robots of the Future..99

9.3 Creating Tomorrow's Workforce...103

CHAPTER 10...106

Using Multipurpose Robots in Your Company 106

    10.1 Planning and Evaluation... 106

    10.2 Strategies for Deployment...111

    10.3 Observation and Ongoing Enhancement....................................116

ABOUT THE AUTHOR.. 120

ACKNOWLEDGMENTS

My sincere appreciation goes out to everyone who helped to make this book possible. I want to start by sincerely thanking the robotics specialists, engineers, and innovators whose work never stops inspiring and pushing the envelope of what is possible. This book would not have been possible without your commitment.

I would especially like to thank my friends, mentors, and coworkers for their crucial advice, criticism, and support during the writing process. Your opinions and knowledge have been extremely helpful in forming the book's content and guaranteeing that it will be applicable to both industry experts and others who are unfamiliar with polyfunctional robotics.

I want to express my gratitude to my family and friends for their constant understanding and support. As I struggled to realize this goal, your encouragement and patience provided me with a solid base.

Lastly, I would like to thank all of the readers of this book for their interest in and dedication to learning more about the future of automation. My goal is to provide you with the information and motivation you need to embrace the fascinating potential of polyfunctional robotics and use it to build a more intelligent and effective future.

I appreciate everyone's participation in this adventure.

CHAPTER 1

Overview of Multipurpose Robots

1.1 Polyfunctional Robots: What Are They?

A major advancement above their task-specific predecessors, polyfunctional robots are a game-changer in the robotics and automation fields. With little assistance from humans, these robots can adapt to dynamic and changing settings and carry out a variety of tasks in a variety of disciplines.

Description and Unique Characteristics

Robots that can execute a variety of activities because of their adaptable hardware and software settings are known as polyfunctional robots. Polyfunctional robots have the following characteristics in contrast to typical robots, which are intended for single, repetitive tasks:

- The ability to rearrange physical components for various applications is made possible by advanced modular designs.

- **Integrated AI systems:** These robots can gradually learn and adjust to new tasks thanks to machine learning and neural networks.
- **Interoperability:** They can operate with ease in a variety of settings and platforms, from hospital rooms to production floors.

They are perfect for handling difficult automation problems because of their capacity to go beyond task-specific constraints.

Robots' Transition from Task-Specific to Multifunctional

The field of robotics has evolved from simple task-specific robots to multipurpose devices with the ability to adapt to new tasks and make decisions in real time. Important turning points include:

- The earliest industrial robots, such as Unimate, were launched in the 1960s and were used for repetitive operations like welding and assembly.
- The 1980s and 1990s saw the introduction of programmable robots that could perform a variety of activities within predetermined constraints.

- The integration of artificial intelligence (AI) and the Internet of Things (IoT) in the 2000s and beyond allowed robots to operate independently in unstructured contexts.

Polyfunctional robots are the apex of this evolution today, combining the greatest features of their forebears with state-of-the-art technology to achieve previously unheard-of versatility.

Value in Contemporary Automation

For a number of reasons, polyfunctional robots are essential to contemporary automation.

- **Enhanced efficiency**: By effortlessly transitioning between jobs, they minimize downtime.
- **Cost-effectiveness:** Companies avoid having to buy and maintain several specialized robots.
- **Adaptability:** They are able to manage the intricacy of changing situations, including disaster response scenarios or warehouses.

In addition to increasing productivity, they are important for building a robust automation ecosystem that can quickly adjust to societal and technological shifts.

1.2 Polyfunctional Robotics' Main Drivers

A number of market and technological factors are combining to satisfy the demands of modern automation, which is driving the growth of polyfunctional robots.

Progress in Machine Learning and Artificial Intelligence

The foundations of polyfunctional robotics are artificial intelligence (AI) and machine learning (ML). These technologies enable robots to:

- **Learn and adapt:** Over time, robots enhance their performance by examining patterns and experiences.
- **Make decisions in real time:** They can evaluate circumstances and carry out the best possible actions by using algorithms.
- **Achieve contextual understanding:** They are able to successfully interpret and engage with their environment thanks to computer vision and natural language processing (NLP).

A polyfunctional robot in healthcare, for example, may help with surgeries, handle logistics, and monitor patients—all of which are powered by advanced AI algorithms.

Requirement for Automation Flexibility

A flexible automation solution is required in today's industries due to the frequent changes in consumer demands and operational settings. Important elements consist of:

- Robots that can manage product variations without retooling are becoming more and more necessary for businesses.
- Polyfunctional robots are scalable, meaning they can be used in both small-scale and large-scale industrial environments.
- **Decreased downtime:** Their capacity to transition between jobs reduces idle time, which is important in sectors like manufacturing and logistics.

Businesses have made significant investments in creating and implementing polyfunctional robots as a result of this need for flexibility.

Market Trends and Economic Considerations

Because of its operational and financial benefits, multifunctional systems are becoming increasingly popular in the worldwide robotics market.

- **Cost of labor:** Automation is becoming a more appealing choice due to labor shortages and rising wages.
- **Return on investment (ROI):** Polyfunctional robots' multitasking capabilities provide long-term savings that offset their greater initial cost.
- **Customization preference of consumers:** Companies using these robots can produce customized goods more quickly and effectively.

Through investments and programs targeted at developing robotic technologies, governments and corporate organizations are further quickening this trend.

1.3 Comparing with Task-Specific Robots

To fully appreciate their transformational potential, it is essential to comprehend the difference between task-specific and polyfunctional robots.

Versatility Has Advantages Over Specialization

In a number of ways, polyfunctional robots are superior to task-specific robots:

- **Broader application scope:** Multiple specialized units can be replaced by a single polyfunctional robot, which lowers the complexity of the infrastructure as a whole.
- **Adaptability to change:** They flourish in erratic settings where predetermined programming fails, such disaster relief or exploration expeditions.
- **Reduced operating costs:** Companies can save money on acquisition, maintenance, and training costs by combining several features into a single platform.

In agriculture, for instance, a multipurpose robot might plant seeds, keep an eye on crops, and gather produce, doing away with the need for additional equipment.

Traditional Robots' Limitations in Changing Environments

Task-specific robots are effective in static contexts, but because of their inflexible design and limited programming, they frequently perform poorly in dynamic or unstructured environments. Important restrictions consist of:

- **Inflexibility:** To accomplish new duties, they need to be extensively reprogrammed or retrofitted.
- **High downtime:** Because specialized robots can only handle one task at a time, there will be periods of inactivity.
- **Limited scalability:** Replicating robotic units is frequently necessary to scale operations, which raises expenses and complexity.

Polyfunctional robots, on the other hand, are able to switch between activities and use AI-driven decision-making to perform well in a variety of situations.

A new era in robotics is represented by polyfunctional robots, which combine versatility and cutting-edge technology to solve contemporary automation problems. These robots stand out as a game-changing answer as industries demand more flexibility and efficiency, paving the way for a future where automation is not only smarter but also more sensitive to the constantly shifting requirements of people.

CHAPTER 2

Polyfunctional Robots' Technological Underpinnings

2.1 Essential Elements

For polyfunctional robots to be flexible and versatile, a complex hardware and software combination is required. Each part is essential to these robots' ability to operate well in a variety of settings and jobs.

Actions and Sensors for Flexibility

The essential parts that allow polyfunctional robots to sense their surroundings and respond with appropriate physical movements are sensors and actuators. By bridging the gap between the digital and physical worlds, these components enable the robot to perform accurate motions and make well-informed decisions.

Sensors:

- **Vision Sensors:** LiDAR, infrared sensors, and cameras offer depth perception and real-time imaging. For jobs like object detection, navigation, and manufacturing quality control, these are essential.
- **Tactile Sensors:** Robots can perform delicate jobs like assembling delicate parts or helping with medical operations thanks to pressure and force sensors.
- Environmental sensors allow robots to operate in dangerous or changeable environments by measuring variables like temperature, humidity, and gas composition.
- **Proximity Sensors:** These aid in preventing collisions and allow for safe mobility in dynamic settings, like hospitals or warehouses.

Actuators:
- **Electric Motors:** These power quick, accurate motions, which are necessary for jobs requiring accuracy and speed.
- **Hydraulic Actuators:** These are ideal for heavy-duty applications and supply the necessary

power for construction and industrial robots.
- Pneumatic actuators are frequently used in settings like food processing and pharmaceutical manufacturing that demand precise and fast movements.

Polyfunctional robots may change between activities with little reconfiguration thanks to the combination of sensors and actuators, which enable dynamic adaptation to their environment.

Frameworks for Software That Facilitate Multifunctionality

The software frameworks of polyfunctional robots act as their brains, allowing them to transition between activities, pick up new skills, and adjust to changing environments. Usually, these frameworks consist of:

- The modular structure of Robot Operating Systems (ROS) allows for the addition or removal of various functionalities as needed. It is a fundamental component of polyfunctional robotics because it facilitates real-time processing, low-level control, and hardware abstraction.

- The integration of machine learning libraries, such as TensorFlow, PyTorch, and Scikit-Learn, enables robots to learn from data and gradually enhance their performance.
- Before being deployed in the real world, robot behaviors can be tested and improved in virtual settings with the aid of software such as Gazebo and V-REP.
- Middleware for Interoperability: By enabling communication between polyfunctional robots and other devices, systems, and cloud services, middleware platforms guarantee smooth data transfer and coordinated actions.

These software frameworks give robots the flexibility needed for multifunctionality by ensuring that they can modify their behavior according to the task at hand.

Design of Modular Hardware

The foundation of polyfunctional robots is modular

hardware design, which makes it simple to replace or reconfigure physical components. This method improves robotic systems' scalability and flexibility.

Interchangeable Modules:
- **End-Effectors:** Depending on the work, tools such as syringes, welding heads, or grippers can be replaced.
- **Mobility Platforms:** To adjust to various terrains, robots can alternate between wheels, tracks, or legs.
- Modules for sensors: Depending on the robot's operational requirements, certain sensors can be added or removed.

Quick-Change Mechanisms:
- Quick-connect systems, which can be mechanical or magnetic, enable operators to quickly switch components, reducing downtime and boosting productivity.

Scalable Architecture:
- By adding or deleting modules, robots can be scaled up or down to perform jobs ranging from complex

assembly to extensive construction.

Because of their modularity, polyfunctional robots can be modified to perform a wide range of jobs without requiring completely new systems, providing flexible and affordable automation solutions.

2.2 Machine Learning and AI's Role

Machine learning (ML) and artificial intelligence (AI) play a key role in converting polyfunctional robots from inflexible, task-specific devices into intelligent, flexible systems. These technologies enable natural human-robot interaction, decision-making, and experience-based learning.

Using Reinforcement Learning to Optimize Tasks

Robots learn through interaction with their surroundings and feedback in the form of rewards or penalties in a process known as reinforcement learning (RL). Polyfunctional robots can optimize their performance for particular jobs through this iterative process.

With the use of adaptive learning, robots can gradually increase their efficiency by modifying their tactics in response to real-world experiences. A robot in a warehouse, for instance, might improve its pathfinding technique to steer clear of traffic.

- **Dynamic Problem-Solving:** Robots can now manage unforeseen tasks like negotiating obstacles or adjusting to changes in product manufacturing lines thanks to reinforcement learning.
- **Continuous Improvement**: A robot's ability to do tasks improves with practice, lowering the requirement for manual reprogramming.

Polyfunctional robots can continuously improve their capabilities and streamline their processes by utilizing reinforcement learning.

Natural Language Processing for Interaction with Humans

Robots can comprehend and react to human voice or text thanks to Natural Language Processing (NLP), which facilitates smooth human-robot interaction.

Voice Commands:
- NLP-enabled robots can carry out activities in response to spoken directions, enhancing usability in settings where hands-free operation is essential, such as healthcare.

Contextual Understanding:
- NLP makes it possible for robots to comprehend tone, intent, and context, facilitating more natural interactions. A polyfunctional robot, for instance, can help consumers in a customer service environment by directing them to particular areas or responding to their inquiries.
- The ability of robots to communicate with users in multiple languages increases their usefulness in a variety of international contexts.

Polyfunctional robots are useful collaborators in collaborative environments because NLP enables efficient communication.

Analysis of Predictive Data for Upkeep

By using artificial intelligence (AI) to predict possible problems before they arise, predictive analytics makes sure polyfunctional robots have as little downtime as possible.

Condition Monitoring:
- Sensors continuously gather information about robot performance, including speed, vibration, and motor temperature.

Failure Prediction:
- Algorithms use this data to forecast component failure times, enabling proactive maintenance scheduling.
- **Cost Savings**: Predictive maintenance lowers repair expenses and downtime by averting unplanned malfunctions.

Predictive analytics improves polyfunctional robots' longevity and dependability while guaranteeing constant performance in difficult settings.

2.3 Cloud computing and IoT integration

New levels of connectivity, data processing, and efficiency are made possible by the combination of polyfunctional robots with cloud computing and the Internet of Things (IoT).

Real-time updates via connectivity

Real-time updates and communication are made possible by IoT technology, which links robots to a network of systems, sensors, and devices.

Remote Monitoring:
Operators may keep an eye on the state and performance of the robots from any location, enabling prompt action in the event of problems.

Dynamic work Allocation:
- IoT allows robots to get real-time work updates in settings like warehouses, such as picking orders based on demand.
- The ability for several robots to cooperate, share information, and modify their activities to maximize efficiency is known as fleet coordination.

Polyfunctional robots are kept nimble and sensitive to demands in the actual world because to this connectedness.

Capabilities for Data Processing and Sharing

Cloud computing gives polyfunctional robots access to enormous computational resources, which improves their data-processing skills.

Centralized Data Storage:
- Data can be uploaded by robots to the cloud, where it can be shared with other systems, stored, and analyzed.
- The use of on-board computing power is decreased by cloud-based AI models that can process complicated tasks and provide instructions back to the robot. This is known as distributed intelligence.
- **Site-to-Site Collaboration:** Robots in various places can exchange updates and insights, fostering network-wide continuous improvement.

Polyfunctional robots can use collective intelligence through cloud integration, which increases their intelligence and productivity.

Improving Productivity With Networked Systems

Polyfunctional robots are more responsive and efficient when IoT and cloud computing are combined.

Automated Workflows:
- Robots can cooperate with other automated systems, including automated guided vehicles (AGVs) or intelligent conveyors.

Real-Time Data Analysis:
- Workflows can be optimized, bottlenecks can be eliminated, and productivity can be increased with the help of insights from real-time data analysis.
- As activities expand, more robots may be easily added to the current network thanks to its scalability.

Polyfunctional robots are guaranteed to perform at their best because of this networked environment, which can swiftly adjust to shifting conditions and needs.

The flexibility, adaptability, and efficiency required for

contemporary automation are provided by the technological underpinnings of polyfunctional robots, which include core components, AI, machine learning, and IoT/cloud integration. Together, these components enable the development of robots that can perform a variety of jobs, increase productivity, and tackle the ever-changing issues of the future.

CHAPTER 3

INDUSTRY-WIDE APPLICATIONS

Polyfunctional robots are bringing efficiency, diversity, and adaptation to a wide range of sectors. Polyfunctional robots are made to move fluidly between a variety of tasks in dynamic situations, in contrast to task-specific robots that are restricted to single-purpose jobs. Their capacity to adapt enables them to effectively address industry-specific demands, improve efficiency, and take on challenging challenges. The influence of polyfunctional robots on manufacturing, healthcare, and agriculture is examined in this chapter, along with particular uses and advantages in each industry.

3.1 Production

Although the use of robotics has historically been led by manufacturing, the move to polyfunctional robots is opening up new possibilities and efficiency. Polyfunctional robots are vital in contemporary manufacturing because of their capacity to carry out several activities on a single production line, adjust to design modifications, and improve quality assurance procedures.

Adaptable Production Lines

Conventional manufacturing assembly lines are frequently inflexible, necessitating significant downtime and reconfiguration to account for output variations. This problem is solved by polyfunctional robots, which provide the adaptability to carry out a range of assembly jobs without requiring significant alterations.

Adaptable Tooling:
- Polyfunctional robots are capable of quickly switching between jobs thanks to their interchangeable end-effectors, which include screwdrivers, welding tools, and grippers. For instance, in just a few minutes, a robot can transition from assembling electronic parts to tightening screws on mechanical assembly.

Dynamic Reprogramming:

- Robots can be dynamically reprogrammed to perform new jobs thanks to contemporary software frameworks. Manufacturers can create customized products more effectively because of this agility, which also promotes shorter product life cycles.
- Polyfunctional robots have the flexibility to scale their workload in response to demand. Additional robots can be readily added to a production line without requiring extensive retooling if output needs to be increased.

This adaptability enables producers to efficiently satisfy the rising demand for customization and quick changes in production.

Inspection and Quality Control

In manufacturing, maintaining product quality is essential. Polyfunctional robots improve quality control by combining advanced sensor technology with analysis driven by artificial intelligence.

Automated Inspection:

- Robots using thermal sensors, LiDAR, and high-resolution cameras may conduct thorough inspections to find flaws like cracks, misalignments, or irregularities.
- Robots can now examine inspection data in real time, highlighting problems right away and even fixing minor flaws on their own thanks to integrated AI systems.
- Robots provide consistent, objective assessment across thousands of products, guaranteeing uniform quality, in contrast to human inspectors who may become fatigued.

These features guarantee that products fulfill strict quality standards, cut down on waste, and lessen the cost of faults.

Management of Inventory

For industrial facilities to continue operating smoothly, effective inventory management is essential. With automation and real-time tracking, polyfunctional robots simplify this procedure.

Automated Material Handling:
- Robots can move semi-finished goods, finished products, and raw materials across the facility, reducing delays and human error.
- Robots that are outfitted with RFID scanners and vision systems are capable of conducting inventory audits and counts on their own, guaranteeing accuracy and lowering the possibility of stockouts or overstocking.
- **Dynamic Storage Optimization:** AI-driven robots are able to rearrange storage according to production requirements, making sure that items that are utilized frequently are in easy reach.

Manufacturers can lower labor costs, increase productivity, and decrease supply chain logistics inefficiencies by automating inventory management.

3.2 Medical Care

The use of polyfunctional robots is causing a revolutionary change in the healthcare industry. These robots greatly enhance healthcare delivery and results by supporting patient care, automating laboratory tests, and assisting with difficult operations.

Helping with Diagnostics and Surgeries

Because they increase accuracy, lower human error, and improve patient outcomes, polyfunctional robots are essential to contemporary surgical and diagnostic operations.

During minimally invasive procedures, robots such as the da Vinci Surgical System help doctors by making extremely accurate motions. Surgeon-controlled robotic arms and high-definition cameras are used by these devices to carry out delicate procedures.
- Robots are adaptable surgical assistants because they can switch between instruments such as scalpels, sutures, and cauterization tools.
- **Diagnostic Procedures:** Robots that are outfitted with imaging technologies like ultrasound, CT, and MRI aid in very accurate ailment diagnosis. For example, biopsies guided by real-time imaging might be performed by a robot, reducing the need for

invasive treatments.

Polyfunctional robots help in diagnostics and surgery, improving surgical accuracy, speeding up recuperation, and lowering risks.

Caring for the Elderly and Patients

The need for senior care solutions is growing as the world's population ages. Polyfunctional robots are a great help when it comes to giving patients individualized care.

- **Mobility Assistance:** Robots can assist elderly or disabled patients with sitting, walking, and getting out of bed and into wheelchairs.
- The ability of robots to distribute medication on time guarantees that patients follow their treatment regimens. They can also keep an eye out for negative responses and urge patients to take their prescriptions.
- **Companion Robots:** Social robots provide company, lowering feelings of isolation and promoting mental wellness. These robots are capable

of cognitive stimulation, gaming, and conversation.

When used in patient care, polyfunctional robots increase independence, improve quality of life, and lighten the effort for medical personnel.

Automation in the Lab

Polyfunctional robots' automation capabilities greatly aid laboratories by improving accuracy, increasing throughput, and streamlining procedures.

Sorting, labeling, and analyzing samples are among the jobs that robots can perform. Their accuracy lowers the possibility of error and contamination.

- **High-Throughput Testing:** Robots are able to conduct thousands of tests at once in domains such as pathology and genomics, which expedites research and diagnosis.
- Laboratory information systems (LIS) can be interfaced with by polyfunctional robots, guaranteeing smooth data reporting and recording.

Lab automation promotes quicker, more accurate diagnostic results, increases efficiency, and lowers human error.

3.3 Farming

Polyfunctional robots are driving a technological revolution in agriculture. These robots solve issues like labor shortages and food security by improving farming operations' efficiency, accuracy, and sustainability.

Farming with Precision

Polyfunctional robots are at the core of precision farming, which uses technology and data to optimize agricultural methods.

Soil Analysis:
- Sensor-equipped robots examine the nutrient content, moisture content, and composition of the soil to help farmers better administer irrigation and fertilizer.
- In order to maximize crop yields while conserving

resources, automated planting robots can plant seeds at the ideal depths and intervals.
- The need for broad-spectrum pesticides is decreased when robots are able to detect pest infestations early and administer customized treatments.

Precision farming boosts yields, encourages sustainable agriculture, and makes better use of resources.

Crop Harvesting and Monitoring

Polyfunctional robots are excellent at keeping an eye on crops and carrying out harvesting duties quickly and efficiently.

- Crop health, growth rates, and disease indicators are tracked by drones and ground-based robots outfitted with imaging sensors. Farmers can use this information to make well-informed decisions.
- **Autonomous Harvesting:** Precision fruit and vegetable harvesting is possible with robots equipped with robotic arms and computer vision, minimizing harm to produce. By changing out their

end-effectors, these robots may be adapted to different crops.

Crop monitoring and harvesting that is automated increases output, lowers labor expenses, and minimizes losses after harvest.

Management of Livestock

Additionally, by automating repetitive duties and enhancing animal care, robots are revolutionizing livestock management.

- **Feeding Automation:** Robots are able to precisely feed animals, guaranteeing optimum nutrition and reducing waste.
- **Health Monitoring:** Robotic sensors can keep an eye on animal health, identifying early indicators of disease or damage. This makes it possible for quick veterinarian care.
- **Milking Systems:** Robotic milking systems adjust to each cow, increasing productivity and milk production while lowering animal stress.

In livestock management, multipurpose robots improve output, protect animal welfare, and simplify agricultural procedures.

Because they provide flexible, accurate, and efficient solutions, polyfunctional robots are revolutionizing the manufacturing, healthcare, and agricultural industries. When paired with cutting-edge AI and sensor technologies, their ability to switch duties smoothly helps enterprises solve contemporary issues, increase productivity, and cut costs. Polyfunctional robots will become more and more important in every industry as these technologies advance, spurring efficiency and creativity.

CHAPTER 4

Polyfunctional Robot Design Principles

Polyfunctional robots' efficiency, versatility, and smooth integration into a range of industries are largely due to their design principles. These robots must be designed with modularity, ergonomics, and sustainability in mind in order to attain versatility and optimize efficacy. These core design concepts will be thoroughly examined in this chapter, offering insights into how each element improves performance and guarantees the best possible deployment in practical settings.

4.1 Scalability and Modularity

Creating Robots for a Variety of Uses

Polyfunctional robots are designed to carry out a variety of activities in a variety of industries. Their design uses modular systems that may be adjusted to meet different needs in order to provide this flexibility. Standardized components or modules that can be mixed or replaced according to the precise function required make up a modular design. This method lowers downtime and expenses related to task-specific robots by enabling rapid and effective reconfiguration of robots.

Standardized Core Units:
- A central processing unit (CPU), sensors, control systems, and motors are usually found in a core robotic platform. While peripheral modules can be altered to suit the requirements of a certain work, the core stays the same.

Reconfigurable Joints and End-Effectors:
- Various jobs call for various tools or accessories. A robotic arm, for instance, can alternate between paint sprayers, welding torches, and grippers. This transformation takes only a few minutes thanks to modular joints and quick-change couplings.

Software Flexibility:
- Modular design encompasses software in addition to hardware. Robots can flip between roles with little reprogramming because they are programmed with

flexible frameworks that can load various task-specific algorithms.

Using Modular Attachments to Increase Functionality

Modular attachments that are added or deleted as needed further increase the adaptability of polyfunctional robots. The robot's capabilities are increased by these accessories, which qualify it for challenging or dynamic jobs.

Attachment Types:
- **Grippers:** For picking and positioning items in logistics, manufacturing, or medical settings.
- **Sensors:** For data collecting, quality assurance, or environmental scanning.
- **Drills or Welders:** For manufacturing assembly line work.
- For jobs involving surveillance, diagnostics, or inspections, use cameras and imaging systems.

The following are quick attachment mechanisms: Quick-lock mechanisms built into contemporary polyfunctional robots enable operators to switch attachments without the need for specialized equipment or intensive training. This improves operating efficiency and decreases downtime.

Adaptability to Various Environments:
- For instance, in agriculture, a robot may utilize a harvesting tool in the afternoon and a soil sensor attachment in the morning. Industries may make the most of a single robotic platform by adapting it to various jobs and situations.

Design Scalability

Scalability is the ability of a system to grow in functionality or capacity without requiring a total redesign. Scalability in polyfunctional robotics guarantees that the same robot can do jobs with different levels of complexity or scope.

Hardware Scalability:
- Robots can be built with modular frames that enable the attachment of further parts. For longer operations, a mobile robot, for example, can be expanded with more battery packs or sensor arrays.
- The ability of AI frameworks and software platforms

to manage growing volumes of data or more complicated activities without seeing a decline in performance is known as "software scalability."
- In settings like factories or warehouses, several robots can work together to create a scalable network that can expand to meet changing operational requirements.

Polyfunctional robots continue to be affordable, flexible, and future-proof thanks to the concepts of modularity and scalability, which satisfy the needs of quickly changing sectors.

4.2 Human-Robot Interaction Ergonomics

Interfaces that are easy to use

Because polyfunctional robots frequently work closely with human employees, smooth interaction depends on having user-friendly interfaces. These interfaces minimize the learning curve and increase productivity by allowing operators to program, monitor, and control robots with little complexity.

Graphical User Interfaces (GUIs):

- Operators may view tasks, configure parameters, and monitor performance in real time with the help of modern robots' user-friendly touch-screen interfaces.

Voice Command Integration:

- Natural language processing (NLP) is used by sophisticated robots to comprehend voice commands, enabling hands-free operation. In industries where operators may need to multitask, like healthcare or logistics, this is very helpful.

Augmented Reality (AR) Interfaces:

- AR overlays on wearable technology, such as smart glasses, offer real-time data and task advice. Without taking their focus away from their work, operators are able to see robotic operations and get directions.

Providing Operator Comfort and Safety

When creating polyfunctional robots for collaborative settings, comfort and safety are crucial considerations. Robots must be designed to function without endangering human employees.

Safety Features:
- **Collision Avoidance Systems:** Robots can modify their movements or halt instantly to prevent collisions when sensors like LiDAR, ultrasonic, and infrared identify human presence and impediments.
- **Emergency Stop Mechanisms:** Robots have readily available emergency stop buttons or sensors that immediately cease working in the event that a safety violation is identified.
- A Soft Robotics System: Soft robotics technology uses malleable materials to lower the risk of damage in settings that require close human-robot interaction.
- The ergonomics of humans must be taken into consideration when designing robots for collaborative tasks. This includes controls that are easy to understand, interfaces that may be adjusted in height, and minimum physical strain on operators.
- Robots help with repeated or heavy lifting jobs in manufacturing, which lowers the risk of musculoskeletal problems in human workers.

Polyfunctional robots increase workplace productivity while preserving a secure, comfortable environment for human collaborators by putting ergonomics and safety first.

4.3 Sustainability and Energy Efficiency

Reducing the Use of Power

For polyfunctional robots to perform sustainably and affordably, energy economy is a crucial design element. In addition to lowering operating expenses, cutting electricity use also advances more general environmental objectives.

Robots are equipped with energy-efficient CPUs, sensors, and actuators that use little power while providing excellent performance.

Energy-Efficient Motion Planning:
- Sophisticated algorithms maximize robot motions while reducing energy use. For instance, robots minimize needless motion and power consumption by planning the quickest or least-resistive routes to accomplish jobs.

Regenerative Systems:

- To increase overall efficiency, certain robots employ motion or braking systems that capture and reuse energy during deceleration.

Using environmentally friendly materials

The materials utilized to construct polyfunctional robots are also sustainable. Eco-friendly products lessen their negative effects on the environment and help businesses meet sustainability goals.

Recyclable Components:
- Robots made of recyclable polymers and metals cut waste and encourage the circular economy.
- At the conclusion of the robot's career, parts including eco-friendly composites, biodegradable polymers, and aluminum frames can be recycled.
- The use of lightweight materials, such carbon fiber, improves efficiency by lowering the energy needed for mobility and transportation.

Non-Toxic Materials:
- Robots in sectors such as food processing and healthcare must be constructed from non-toxic,

food-safe materials that adhere to safety and health regulations.

Reduction of Environmental Impact

By lessening the environmental impact of industrial processes, polyfunctional robots help achieve more general sustainability objectives.

Waste Reduction:
- Robots in manufacturing use precision operations to decrease waste and maximize resource consumption.
- Robots that are equipped with renewable energy sources, such solar-powered charging stations, further lessen dependency on non-renewable resources.

Sustainable Lifecycle Management:
- Robots with modular and upgradeable designs last longer, requiring fewer replacements and producing less e-waste.

Polyfunctional robots meet current industry standards for cost-effectiveness and environmental responsibility by emphasizing energy efficiency and sustainable materials.

The success of polyfunctional robots in dynamic situations is largely dependent on their design concepts, which include modularity and scalability, ergonomics for human-robot interaction, and energy efficiency and sustainability. These guidelines guarantee that while achieving optimal performance, robots continue to be safe, flexible, and ecologically benign. These design factors will become more crucial in determining how automation and human-robot cooperation develop in the future as industries continue to change.

CHAPTER 5

Polyfunctional Robotics Challenges

Polyfunctional robots, which provide flexibility, efficiency, and adaptation across a range of industries, represent a significant leap in automation and technical innovation. Nevertheless, despite their potential, there are a number of obstacles that must be overcome before these robots can be widely used. These difficulties can be divided into three main groups: technological difficulties, financial and ROI concerns, and moral and legal dilemmas. To maximize the potential of polyfunctional robotics, these issues must be recognized and resolved.

5.1 Technical Difficulties

Overcoming Complexity in Programming

The intricacy of programming these systems is one of the biggest obstacles in polyfunctional robotics. Polyfunctional robots are required to manage a variety of jobs, frequently in dynamic and unpredictable contexts, in contrast to task-specific robots, which are made to carry out a single, repetitive function. This degree of adaptability necessitates sophisticated software frameworks and extensive programming skills.

The following are adaptive algorithms:

Polyfunctional robots use adaptive algorithms that can change their behavior in response to shifting circumstances in order to attain multifunctionality. Real-time decision-making, reinforcement learning, and machine learning (ML) models must all be integrated for this to work. These algorithms are difficult to develop and require a high level of robotics and artificial intelligence (AI) expertise.

Task switching and multitasking:

Robots that are polyfunctional must be able to transition between activities with ease, such as assembling parts and doing quality checks. Programming logic, sensor inputs, and control techniques may vary depending on the task. One of the biggest programming challenges is making sure that these transitions happen error-free and seamlessly.

Friendly User Interfaces:

Even though polyfunctional robots require complex backend programming, the user interface must be simple enough for operators without extensive technical expertise to understand. For developers, creating interfaces that streamline intricate processes without compromising control or flexibility is a never-ending task.

The process of debugging and maintaining

Debugging and maintaining polyfunctional robots is made more challenging by their complexity. Finding the source of problems and putting a solution in place can take a lot of effort and require certain knowledge.

Reliability and Versatility in Balance

Although polyfunctional robots are made to do a variety of jobs, their adaptability occasionally compromises their dependability. Polyfunctional robots have to balance flexibility and reliable performance, in contrast to specialized robots designed for a specific purpose.

Mechanical Damage and Wear:

The robot may experience differing degrees of wear and tear on different sections due to its capacity to transition between different duties. For instance, a robotic arm that performs both heavy lifting and delicate assembling would degrade mechanically more quickly. Reliability requires the design of robust components that can handle a variety of jobs.

Sensor Calibration and Accuracy:

distinct kinds of sensors, such cameras, force sensors, or temperature sensors, are frequently needed for distinct activities. It is technically challenging to guarantee that all sensors maintain their precise calibration when alternating between jobs. Errors caused by inaccurate sensor readings can lower the robot's dependability.

The software's stability

Bugs and malfunctions are common in multifunctional software systems, especially when alternating between intricate tasks. Maintaining software dependability requires making sure it stays responsive and stable in a variety of scenarios. Error management, thorough testing, and real-time monitoring systems are necessary for this.

The constant task of striking a balance between adaptability and dependability necessitates constant innovation in software development, sensor technologies, and hardware design.

5.2 ROI and Cost Factors

High Initial Outlay

Polyfunctional robot deployment requires a large upfront expenditure. Because these robots are more advanced and complex than task-specific robots, their development, hardware, software, and integration expenses are higher. The price is further increased by the requirement for sophisticated sensors, AI algorithms, and modular components.

Cost Drivers:
- **Hardware Costs:** Modular attachments, high-quality sensors, and actuators are costly.

- **Software Development:** It takes a lot of effort and experience to create machine learning and adaptable AI models.
- **Training and Implementation:** To operate with polyfunctional robots, operators and technicians require specific training, which raises the total cost.

The infrastructure needs to be as follows:

It is frequently necessary to modify current infrastructure in order to implement polyfunctional robots. For instance, it may be expensive to integrate these robots with cloud computing, data analytics platforms, and Internet of Things (IoT) networks.

Cost-Reduction Techniques

Several tactics can be used to lower the total investment and increase return on investment in order to make polyfunctional robots more affordable and accessible.

Design Modularity:

Businesses can increase functionality over time without investing in whole new systems by investing in modular robots. This method lessens the requirement for significant initial outlays.

Models of Leasing and Subscription:
By implementing robotics leasing or subscription models, businesses can lower their capital expenditure. Without needing significant upfront expenditures, these approaches offer access to the newest technology and distribute costs over time.

Software that is open-source:
Development costs can be decreased by utilizing open-source robotics software. Strong solutions that can be tailored for particular needs are frequently offered by communities and cooperative ventures.

Cobots, or collaborative robotics:
Compared to completely autonomous systems, collaborative robots that are made to work with people may be less expensive. Generally speaking, cobots are simpler, smaller, and simpler to incorporate into current workflows.

The ability to scale and deploy incrementally:

Before making more significant investments, companies can evaluate and improve processes by implementing polyfunctional robots in phases. This incremental strategy allows for steady scaling while lowering financial risk.

Determining ROI

Organizations must carefully assess ROI in order to justify the investment in polyfunctional robots. Important things to think about are:

- Redirecting human resources to higher-value tasks and reducing manual labor are two ways to save labor costs.
- Faster production cycles, better quality control, and less downtime are all examples of increased efficiency.
- The ability to transition between tasks without requiring further capital investment is one of the flexibility gains.
- Decreased Errors: Automation lowers the possibility of human error, resulting in output of superior quality.

Businesses can make well-informed judgments about the use of polyfunctional robotics by carefully evaluating these criteria.

5.3 Concerns about Regulation and Ethics

Observance of Industry Standards

For polyfunctional robots to be used safely, dependably, and ethically, they must adhere to a number of industry standards and laws. The industry, geography, and application all affect these criteria.

- **ISO 10218:** Safety criteria for industrial robots and robot systems are outlined in the following safety standards.
- **ANSI/RIA R15.06**: American guidelines for the safety of robots in manufacturing settings.
- Functional safety requirements for electronic and programmable systems are outlined in IEC 61508.

The regulations pertaining to data privacy:

Robots using sensors and artificial intelligence (AI) systems frequently gather vast volumes of data. Sensitive information must be protected by adherence to data privacy laws, such as the General Data Protection Regulation (GDPR) in Europe.

Regulations Concerning the Environment:
- Regulations controlling emissions, waste management, and the usage of hazardous materials must all be adhered to by sustainable design methods.

Continuous audits, documentation, and monitoring are necessary to guarantee adherence to these requirements.

Resolving Job Displacement Issues

The possibility of employment displacement is one of the biggest ethical issues with polyfunctional robotics. There is fear that human workers may be displaced by robots as they grow more competent and adaptable, which would cause social and economic upheaval.

The process of reskilling and upskilling

By funding training initiatives that assist employees in acquiring new skills, organizations can lessen the impact of job displacement. This enables workers to move into professions like robot maintenance, programming, or supervision that enhance robotic automation.

Robot-Human Cooperation:

A balance between automation and human work is maintained by placing a strong emphasis on collaborative robots, or cobots. Cobots increase efficiency without displacing human workers; they complement them.

Frameworks for Ethics:

To guarantee that automation benefits society overall, ethical standards for robotics deployment should be developed. This entails encouraging ethical work practices, helping impacted employees, and taking the social effects of automated decisions into account.

To ensure that the adoption of polyfunctional robotics results in inclusive and sustainable growth, industry leaders, legislators, and society must work together to address these ethical issues.

Although there are still many obstacles to overcome, polyfunctional robotics has the potential to revolutionize several industries. Achieving successful implementation requires overcoming programming complexity, striking a balance between adaptability and dependability, controlling expenses, and attending to ethical and legal issues. Organizations can fully utilize polyfunctional robots while guaranteeing ethical, economical, and sustainable results by carefully and strategically tackling these issues.

CHAPTER 6

COOPERATION BETWEEN HUMANS AND ROBOTS

A major advancement in automation has been made with the emergence of polyfunctional robots, which will usher in a time when people and robots can work together harmoniously in a variety of industries. The goal of human-robot collaboration (HRC) is to increase job happiness, safety, and productivity by utilizing the skills of both humans and machines. This cooperation is especially beneficial for polyfunctional robots because of their diversity, adaptability, and capacity to carry out a variety of activities in tandem with humans. Effective collaboration, however, necessitates focus on staff training, safety regulations, and efficiency improvement. These aspects are thoroughly examined in this chapter, offering a thorough grasp of the potential and difficulties in human-robot cooperation.

6.1 Polyfunctional Robots' Function in Collaborative Work Environments

The purpose of polyfunctional robots is to function in dynamic, cooperative settings. Because of their versatility, they are ideal for sectors including manufacturing, logistics, healthcare, and agriculture that demand flexibility. Polyfunctional robots complement human workers in collaborative workspaces, increasing productivity and allowing for more strategic use of human labor.

Improving Efficiency and Productivity

The special qualities of polyfunctional robots, such as accuracy, speed, and consistency, are advantageous for collaborative workspaces. When used efficiently, these robots are able to:

Automate Tasks That Are Repeated:
Robots can be used to perform repetitive and boring jobs like data entry, packaging, and assembly. This frees up human workers to concentrate on more intricate,

imaginative, and decision-making activities that call for emotional intelligence, problem-solving skills, and cognitive capacities.

Conduct Several Tasks:

Polyfunctional robots are capable of switching between duties as needed, in contrast to typical robots that are restricted to a particular purpose. For instance, a single robot may do inventory management, assembly, and quality control in a manufacturing setting, negating the need for several specialized machines.

Enhance Workflow Flexibility: Collaboration between humans and robots enables workflows to adjust to shifting circumstances. Polyfunctional robots can be reprogrammed or reconfigured to perform extra activities in the event of an increase in manufacturing demand, guaranteeing that productivity stays high.

Decrease Downtime:

Robots can continue to work in collaborative settings when human workers are not present, for example during breaks or shift changes. Productivity is increased and downtime is reduced by this ongoing activity.

Cutting Down on Human Labor in Repeated Tasks

The decrease of human burden in cognitively and physically taxing jobs is one of the main advantages of polyfunctional robots in collaborative workspaces. There are several benefits to this reduction:

The reduction of fatigue:
Physical exhaustion and repetitive strain injuries (RSIs) can result from repetitive activity. Employers can lower accident rates and enhance worker well-being by delegating certain activities to robots.

Enhancing Job Satisfaction and Morale:
Employees can work on more rewarding and intellectually challenging projects when they are released from repetitive duties. Increased job satisfaction, better morale, and lower turnover rates can result from this change.

Enhancing Human Capabilities:
Organizations can better utilize human abilities with the

help of polyfunctional robots. While robots take care of the repetitive chores, workers may concentrate on jobs that call for abilities like creativity, problem-solving, and interacting with customers.

Polyfunctional robots make the workplace more effective and balanced for both humans and machines by increasing productivity and decreasing workload.

6.2 Safeguarding Human-Robot Communication

In settings where people and robots coexist, safety is crucial. In collaborative environments, polyfunctional robots must cohabit with humans in shared workstations, in contrast to traditional robots that frequently work in separate locations. Advanced technologies, adherence to safety regulations, and continual risk assessment are necessary to ensure safe interactions.

Complex Sensing and Reaction Systems

With the help of sophisticated sensor technologies, polyfunctional robots can recognize and react to human presence. Among these technologies are:

Proximity sensors are:
In order to prevent collisions, proximity sensors sense when a human is close to the robot and cause it to slow down, stop, or alter its path.

Depth sensing and computer vision:
LiDAR (Light Detection and Ranging) and camera-equipped robots are able to sense their surroundings in real time. Robots can detect impediments, understand human gestures, and modify their movements in response thanks to these sensors.

Sensors of Force and Torque:
Robots can detect the force they are delivering with the aid of these sensors. The robot can halt right away to avoid harm if it detects unexpected opposition.

Stop Systems for Emergencies:
In order to immediately stop activities in the event of danger, robots are built with emergency stop mechanisms that can be activated by automated systems or humans.

In order to reduce hazards and guarantee that polyfunctional robots can perform securely in cooperative settings, these sensing and response systems are essential.

Guidelines for Secure Co-Working Spaces

Polyfunctional robots must abide by a number of industry norms and laws to guarantee safety. Guidelines for the development, implementation, and use of robots in human-centered settings are provided by these standards:

ISO 10218:

The safety standards for industrial robots and robot systems are outlined in this international standard. It covers safety functions, protection measures, and risk assessment.

ANSI/RIA R15.06:

Guidelines for robot safety in manufacturing settings are provided by the Robotics Industries Association (RIA) and the American National Standards Institute (ANSI). These guidelines place a strong emphasis on protective barriers and risk assessment.

ISO/TS 15066 Collaborative Robot Standards:

The safety criteria for collaborative robots (cobots) are the main topic of this technical specification. It describes the fundamentals of safe contact, such as force and speed restrictions and safety precautions.

CE Marking:

Robots in Europe must be CE marked in order to prove that they adhere to safety, health, and environmental regulations.

For human-robot collaborative workspaces to be safe, effective, and compliant, adherence to these standards is crucial.

6.3 Adoption and Training

A workforce that is knowledgeable, flexible, and comfortable with these technologies is necessary for integrating polyfunctional robots into collaborative workspaces. Upskilling staff members and creating a welcoming and trustworthy workplace are essential for a successful adoption.

Employee Upskilling for Smooth Integration

Employees frequently need to acquire new skills and abilities when polyfunctional robots are introduced. Workers can adjust to these changes with the aid of upskilling programs:

Training for Robot Operation and Maintenance:

Employees must be proficient in the use, maintenance, and troubleshooting of polyfunctional robots. Basic programming, safety protocols, and robot interfaces are all covered in this class.

Skills in Automation and Programming:

Workers with rudimentary programming and automation skills will be in high demand as robots get more complex. Workers can be empowered to participate in robot deployment through training programs in robotics programming languages like ROS (Robot Operating System) or Python.

Skills in Data Analysis:

Large volumes of data are frequently generated by polyfunctional robots. Employers can gain insights and improve robot performance by teaching employees how to analyze and understand data.

Risk and Safety Management:

To guarantee safe cooperation with robots, workers need to get training on safety measures, risk assessments, and emergency protocols.

Promoting Acceptance via Instruction

Addressing possible opposition to change and promoting an accepting culture are crucial for a successful adoption. Among the tactics to promote acceptance are:

Transparent Communication:

Clearly explain the advantages of polyfunctional robots, such as how they will increase output, lighten workloads, and open up new career paths for employees.

Examinations and Practical Experience:

Encouraging employees to engage with robots in a safe setting demystifies the technology and fosters trust.

Participate in the integration process by involving staff members in the design and implementation of multipurpose robots. Their suggestions might assist in modifying the integration procedure to take into account demands and issues from the actual world.

Ongoing Assistance and Input:
Continue to offer assistance, instruction, and feedback opportunities. Improving acceptance and trust can be achieved by immediately addressing issues and taking suggestions into consideration.

Organizations can guarantee a seamless transition to collaborative workspaces where people and multifunctional robots collaborate efficiently by making training and educational investments.

The future workplace is built on human-robot collaboration, with polyfunctional robots increasing output, decreasing workload, and fostering more dynamic, secure settings. Addressing productivity targets, putting strong safety measures in place, and funding personnel training are all necessary to ensure productive teamwork. Organizations can achieve unprecedented levels of productivity and creativity while cultivating a flexible and inclusive workplace culture by carefully incorporating polyfunctional robots.

CHAPTER 7

Polyfunctional Robots' Economic Impact

With their adaptability, efficiency, and diversity across a range of applications, polyfunctional robots are bringing about a revolutionary change in the way industries function. These robots have significant economic ramifications in addition to their technological benefits. firms looking to implement these technologies must comprehend their cost-effectiveness, employment impact, and potential for small and medium-sized firms (SMEs). This chapter explores the economic environment that polyfunctional robots have shaped, offering a thorough examination of their cost-benefit ratios, the changing nature of the workforce, and their capacity to strengthen SMEs.

7.1 Analysis of Cost-Benefit

Organizations thinking about implementing polyfunctional

robots must conduct a thorough cost-benefit analysis. The study assists in calculating the return on investment (ROI) across different industries and comparing these sophisticated systems with conventional robots.

A Comparison of Polyfunctional and Conventional Robots

Conventional Robots:

Conventional industrial robots are usually made to perform a specific, specialized operation, like packaging, welding, or assembly. These robots are very efficient and precise, but they are not very flexible. Their implementation frequently necessitates a large infrastructural, programming, and maintenance expenditure. Costly reprogramming or hardware changes are required whenever production requirements change, such as when the design of the product changes. These robots' fixed-function design restricts their use and flexibility in dynamic sectors.

Robotics with many functions:

Conversely, polyfunctional robots are made to do a variety

of jobs. These robots can adapt to various tasks without requiring significant redesigns since they are outfitted with modular parts, sophisticated sensors, and intelligent software. The following are the main advantages of polyfunctional robots:

They are perfect for industries with changing demands because of their flexibility, which allows them to be swiftly reprogrammed and reconfigured to carry out new tasks.

- **Decreased Downtime**: Changing jobs can be completed with less downtime, which boosts output in general.
- One polyfunctional robot can take the place of several specialized robots, resulting in a reduced total equipment expenditure.
- **Scalability:** The robot's capabilities can be increased with new modules or software upgrades as company requirements change.

The long-term advantages of polyfunctional robots frequently exceed their initial cost, even if they may be more expensive than regular robots.

ROI Schedules by Industry

Polyfunctional robots' return on investment (ROI) varies according to the industry, task complexity, and deployment scale. A summary of average ROI timelines for several businesses is shown below:

The manufacturing process:
ROI can be attained in 1-2 years in high-volume production settings because of improved productivity, lower labor costs, and fewer mistakes. Multiple systems are less necessary when polyfunctional robots are used to manage inventory, assembly, and quality assurance.

Healthcare:
Polyfunctional robots help with patient care, diagnosis, and surgery in healthcare environments. Because medical-grade robotics are expensive and require strict safety regulations, the return on investment (ROI) period is sometimes two to four years. Nonetheless, the expenditure is justified by the increased precision, shorter operating times, and better patient results.

Agriculture:

ROI in crop management and precision farming can be achieved in two to three years. With less human involvement, polyfunctional robots can plant, tend, and harvest crops, increasing yields and reducing labor expenses.

Logistics and Warehousing:

ROI can be attained in 1-2 years in distribution facilities and e-commerce. Picking, packing, and inventory tracking are made easier by polyfunctional robots, which lower operating expenses and expedite order fulfillment.

Long-term financial gains can be obtained by investing in polyfunctional robots, particularly in sectors that need efficiency and flexibility. These robots' adaptability and agility guarantee that businesses can promptly adjust to shifts in the market and in technology.

7.2 Employment Impact

The workforce will be greatly impacted by the emergence of polyfunctional robots. Although there are legitimate worries about job displacement, these robots also open up new career prospects by moving the nature of labor toward more specialized and technical positions.

Changing Work Roles to Include More Technical Knowledge

Human employees can concentrate on higher-level duties by having polyfunctional robots handle labor-intensive, repetitive activities. As a result of this change, new positions requiring technical knowledge are created, including:

To create, manage, and modify robot functions, people with programming expertise in languages like Python, C++, and ROS (Robot Operating System) are required.

The following system integrators:
These experts make sure that robots integrate well with current procedures and systems. They are in charge of integrating data analytics tools, software, and hardware.

Data Interpreters:

Robots with multiple functions produce a lot of data. By interpreting this data, data analysts can increase decision-making, streamline operations, and maximize robot performance.

Technicians for maintenance:
Polyfunctional robots must be maintained and repaired by skilled specialists to ensure their safe and effective operation.

Developing Prospects in Robot Programming and Maintenance

As polyfunctional robots become more widely used, there is a greater need for specialist positions in programming and maintenance. Opportunities for career growth and workforce development are created by this trend:

Reskilling Initiatives:
Employers can fund reskilling initiatives to teach current staff members data analysis, programming, and robotics maintenance. These initiatives guarantee that workers maintain their employability and technical flexibility.

The field of technical education

To train the next generation of workers for professions in robotics, educational institutions can extend or develop robotics and automation courses.

High-Value Jobs: Programming, system integration, and robotics maintenance jobs usually pay more and provide more job stability than manual labor positions.

The creation of new positions guarantees that the workforce can adjust to the changing technology landscape, even while some old jobs may be phased out. Organizations may lessen the negative effects of automation and develop a workforce prepared for the future by emphasizing education and reskilling.

7.3 Strengthening SMEs (small and medium businesses)

Polyfunctional robots aren't limited to big businesses. They provide small and medium-sized firms (SMEs) with a number of benefits that help them compete in the global market, increase productivity, and cut expenses.

Availability of Inexpensive Robotics Products

Technological developments have reduced the cost and increased the accessibility of polyfunctional robots for SMEs. Among the main causes of this tendency are:

Reduced Hardware Expenses:
Polyfunctional robots are now more inexpensive thanks to advancements in technology that have reduced the cost of sensors, actuators, and processing units.

Models of Subscriptions:
SMEs can lease robots instead of buying them altogether thanks to Robots-as-a-Service (RaaS) models offered by certain robotics suppliers. This method offers flexibility and lowers up-front expenses.

Open-Source Software: SMEs may create and modify robot applications without having to pay for expensive software thanks to open-source robotics platforms like ROS.

Design Modularity:

SMEs can begin with simple functions and add more capabilities as needed thanks to polyfunctional robots with modular designs. Smaller companies can make incremental investments because of its scalability.

Strengthening Market Competition

SMEs can more successfully compete with larger businesses thanks to polyfunctional robots by:

- By automating processes like assembly, packing, and quality control, robots help SMEs boost productivity without having to hire a large number of new employees.

- **Improving Quality:** Accurate and consistent operations lower errors and faults, improving customer satisfaction and product quality.

Adjusting to Shifts in the Market:

Polyfunctional robots' adaptability enables SMEs to swiftly respond to shifting consumer needs, new product launches, or industry changes.

Cutting Labor Expenses:
SMEs can save labor expenses and free up human resources for more crucial responsibilities by automating monotonous processes.

By integrating polyfunctional robots, SMEs may level the playing field with larger rivals by increasing productivity, upholding quality standards, and reacting quickly to market opportunities.

Polyfunctional robots have a significant economic impact on employment dynamics, cost structures, and company competitiveness. It is evident from a thorough cost-benefit study that these robots provide long-term financial benefits. Their acceptance ensures that workers stay relevant in an automated environment by reshaping the workforce and opening up new chances for technical professions. Polyfunctional robots' accessibility and adaptability provide SMEs a route to expansion and competitiveness. Businesses of all sizes can set themselves up for a vibrant and effective future by embracing this technology.

CHAPTER 8

MULTIPURPOSE ROBOTS IN INTELLIGENT CITIES

Thanks to developments in automation, the Internet of Things (IoT), and digital infrastructure, the idea of smart cities has rapidly changed in recent years. The incorporation of polyfunctional robots—systems that are adaptable, versatile, and able to efficiently complete several tasks is at the core of this revolution. Cities are becoming more sustainable, efficient, and habitable thanks to polyfunctional robots. This chapter examines how these robots can improve urban living, revolutionize logistics and transportation, and aid in green and sustainable projects.

8.1 Improving City Life

Urban infrastructure is under greater and more pressure to uphold strict standards for public services, safety, and cleanliness as cities get denser. Polyfunctional robots contribute to a cleaner, safer, and more effective urban environment by offering flexible solutions for recycling, trash management, and public safety.

Robots for Recycling and Waste Management

One essential element of smart cities is effective waste management. Traditional waste collection systems generally deal with inefficiencies, pollution, and expensive labor costs. Waste management is revolutionized by polyfunctional robots through:

The ability of autonomous waste-collecting robots to precisely identify and collect rubbish while navigating public areas and city streets is known as "automated waste collection." These robots can differentiate between various trash types and sort them appropriately since they are outfitted with sophisticated sensors and machine learning algorithms.

Ingenious Recycling Options:
Compared to human-operated systems, polyfunctional robots can sort recyclable materials more accurately. These

robots can distinguish between plastics, metals, glass, and paper using computer vision and artificial intelligence, guaranteeing that recycling streams are kept clean. This promotes sustainability objectives and increases the effectiveness of recycling facilities.

Route optimization and dynamic scheduling:

Waste-collection robots can use real-time data to calculate the best routes and collection times by integrating with IoT systems, which lowers emissions and fuel consumption. Additionally, this lessens traffic brought on by garbage collection trucks.

Cleaning and Maintenance:

Polyfunctional robots can be used for sanitation in public spaces, graffiti removal, and street cleaning in addition to rubbish collecting. Because of their versatility, they can handle a variety of tasks, without the need for additional equipment.

Applications for Public Safety and Surveillance

Any smart city must prioritize public safety. In order to

improve security, keep an eye on public areas, and react to crises, polyfunctional robots can be crucial:

Self-governing Patrols:
Public spaces like parks, streets, and transit hubs can be patrolled by robots outfitted with cameras, sensors, and artificial intelligence. These robots are able to recognize possible dangers, spot suspicious conduct, and instantly notify human operators.

Reaction to Emergencies:
Polyfunctional robots can be used in emergency scenarios, including fires, chemical spills, or natural catastrophes, to evaluate damage, look for survivors, and provide assistance. They are crucial to first-response teams because of their capacity to function in dangerous situations.

Crowd Management and Control:
Robots can help with crowd control, real-time information, and directing people to safe areas during major public gatherings. By adjusting to changing circumstances, these robots can minimize human interference while maintaining public safety.

Data collection and surveillance:

Data on air quality, traffic patterns, and public safety indices can be gathered by polyfunctional robots. Informed decisions to enhance urban living circumstances are made by law enforcement and city planners with the use of this data.

Cities become safer, cleaner, and more effective when polyfunctional robots are used in garbage management and public safety. These robots' versatility guarantees that they can handle several urban problems at once.

8.2 Logistics and Transportation

The foundation of any smart city is effective transportation and logistics. With autonomous deliveries, traffic control, and vehicle maintenance, polyfunctional robots are revolutionizing urban transportation systems.

Self-sufficient Delivery Mechanisms

The need for quick, effective delivery methods has grown

as e-commerce and online services have become more popular. Polyfunctional robots provide creative urban logistics solutions:

Delivery of the Last Mile:
Packages can be delivered straight to clients by autonomous delivery robots that can handle bike lanes, sidewalks, and pedestrian zones. To avoid obstructions, choose the best routes, and guarantee on-time delivery, these robots are outfitted with sensors and GPS.

Aerial Delivery Drones:
Delivery drones may move items via the air, avoiding traffic in crowded urban areas. Polyfunctional drones are flexible instruments for urban logistics because they can adjust to various payloads, delivery sites, and weather conditions.

An example of warehouse automation is:
Polyfunctional robots pick, pack, and sort items to expedite warehouse operations in tandem with delivery systems. Their versatility lowers the need for specialist equipment, boosting productivity and cutting expenses.

Vehicle Maintenance and Traffic Control

In many places, traffic congestion is a major problem that causes inefficiencies, delays, and pollution. Polyfunctional robots tackle these issues by:

A smart way to monitor traffic

Real-time traffic flow monitoring by robots with sensors and artificial intelligence (AI) can pinpoint areas of congestion and recommend the best routes for drivers. To enhance traffic management, this data can be combined with intelligent traffic signals and signage.

Vehicle Maintenance and Inspection:

Autonomous robots are capable of performing basic maintenance, identifying mechanical problems, and inspecting vehicles. By keeping delivery trucks, private automobiles, and public transportation fleets in top shape, this lowers breakdown rates and enhances road safety.

Management of Parking:

By directing cars to open spots, enforcing parking laws,

and maximizing space utilization, polyfunctional robots can help with parking space management.

Polyfunctional robots increase the efficiency of urban transportation by improving delivery systems and traffic management, which lowers emissions, congestion, and operating expenses.

8.3 Green initiatives and sustainability

In order to lessen their impact on the environment and encourage green living, smart cities must prioritize sustainability. Polyfunctional robots are essential for circular economies, waste reduction, and renewable energy systems.

Robots in Systems for Renewable Energy

Solar, wind, and hydroelectric power systems are examples of renewable energy infrastructure that is developed and maintained with the help of polyfunctional robots:

Upkeep of Solar Panels:

Solar panels can be cleaned and inspected by robots to guarantee optimal performance. They are essential for sustaining massive solar farms because of their versatility in handling various panel types and weather conditions.

An examination of wind turbines:

Drones with multiple uses can check wind turbines for wear, damage, and inefficiency. These robots can spot problems that human inspectors would miss because they are outfitted with high-resolution cameras and sensors.

Management of the Energy Grid:

Robots are capable of power grid monitoring and maintenance, defect detection, energy distribution optimization, and downtime reduction. Their versatility guarantees the resilience and effectiveness of energy infrastructure.

Encouragement of Circular Economy

The goal of a circular economy is to maximize resources and reduce waste. This model is supported by polyfunctional robots, which improve resource recovery, waste management, and recycling:

Recovery of Materials:

Robots are capable of sorting and recovering valuable materials from domestic garbage, building detritus, and technological waste. This lessens the impact on the environment and the requirement for raw materials.

Repair and Remanufacturing:

Polyfunctional robots can help with re-manufacturing procedures, deconstruct items, and locate reusable parts. This lessens the need for new resources and encourages the reuse of existing ones.

Systems that convert waste into energy:

Robots can optimize trash processing in waste-to-energy facilities to produce power, minimizing landfill usage and fostering the creation of sustainable energy.

Polyfunctional robots help communities meet their sustainability targets, lessen their environmental effect, and foster a greener future by assisting with renewable energy and circular economy efforts.

Polyfunctional robots are at the vanguard of smart city development, offering various solutions to enhance urban living, improve transport and logistics, and assist environmental programs. They are essential to building safer, cleaner, and more effective cities because of their versatility. Polyfunctional robots will play an increasingly important role in smart cities as technology develops, converting urban settings into dynamic, adaptable, and sustainable places for coming generations.

CHAPTER 9

Polyfunctional Robotics' Future Trends

As new technologies redefine the potential of what robots can accomplish, polyfunctional robotics is embarking on an exciting period of innovation and integration. We may anticipate that robots will not only advance in sophistication but also grow more cooperative and adaptive in the years to come. Advances in artificial intelligence, material science, computers, and human-robot interaction will propel this progress. Beyond industry and technology, these developments will have a big impact on the workforce and educational systems around the world.

The main themes that will influence polyfunctional robotics in the future will be discussed in this chapter, along with the technologies that are propelling innovation, the upcoming intelligent robots, and ways society may be ready for a future driven by robotics.

9.1 New Technologies Fueling Creativity

Cutting-edge developments in a number of disciplines, most notably material science and quantum computing, will be advantageous for polyfunctional robots. With the help of these technologies, robots will be able to perform tasks more quickly, intelligently, and robustly. Comprehending these advancements is essential to predicting the robotics environment of the future.

Robotics and Quantum Computing

With its unmatched processing capacity, quantum computing has the potential to completely transform robotics. Quantum computers employ quantum bits (qubits) that can exist in several states at once, in contrast to classical computers that process information in binary (0s and 1s). This makes it possible for quantum computers to execute intricate computations at exponentially faster speeds.

Quantum computing has the potential to spur

innovation in polyfunctional robots in a number of ways:

Pathfinding and Optimization:
In dynamic situations like warehouses or metropolitan landscapes, robots that are performing numerous tasks must swiftly choose the best course of action. These optimization issues can be resolved more quickly by quantum algorithms, which will speed up decision-making and increase robot output.

Improving AI and Machine Learning:
Large amounts of computing power are needed for advanced AI models, especially when they are learning from big datasets. Robots can learn faster and adjust to new tasks more easily thanks to quantum computing's ability to speed up machine learning algorithm training durations.

The process of simulating complex systems:
Simulations are necessary for robots working in uncertain contexts (such as disaster areas or space exploration) in order to forecast results and make plans of action. These simulations can be handled more precisely by quantum

computing, which improves decision-making and reduces errors.

Security and Cryptography:

Quantum encryption can provide tighter security measures, safeguarding sensitive data and guaranteeing the integrity of robotic operations as robots depend more and more on data transmission and communication.

Even though quantum computing is still in its infancy, robotics could benefit greatly from continued study and development. As these developments permeate mainstream robotics, early adopters in sectors like logistics, healthcare, and defense will profit.

Progress in the Sciences of Materials

Another area influencing the development of polyfunctional robotics is material science. Robots can now be used in a greater variety of applications because of new materials that make them stronger, lighter, and more adaptable. Current robotic systems have certain physical restrictions that are being addressed by innovations in this field.

The following major developments in material sciences are propelling innovation in robotics:

Self-Healing Substances:
Self-healing materials that can fix slight damage on their own are being made possible by developments in polymers and nanomaterials. These materials can be used to build robots that have a longer lifespan, require less maintenance, and operate in dangerous conditions with a lower chance of malfunction.

Soft and Flexible Robotics:
Soft robotics, which draws inspiration from biological systems, makes use of flexible materials such as hydrogels, silicone, and elastomers. These materials enable robots to perform sensitive activities including food processing, agriculture, and medical operations. Soft robots' utility is increased by their capacity to adjust to uneven surfaces and small areas.

High-strength, lightweight alloys:

The ideal ratio of strength to weight is offered by new metal alloys and composite materials. These materials are necessary for wearable robotics, exoskeletons, and drones, where durability and weight reduction are critical.

SMAs, or shape-memory alloys, are:
When heated, SMAs can revert to a predetermined shape. Actuators and robotic limbs benefit greatly from this feature since it enables them to execute intricate movements without the need for conventional motors or gears.

Materials Inspired by Biotechnology:
Scientists are using nature as a model to create materials that are as resilient and adaptive as biological tissues. Synthetic spider silk, for instance, has remarkable strength and elasticity, which makes it perfect for robotics applications that need sturdy yet lightweight structures.

These developments will increase the capabilities and adaptability of polyfunctional robots, opening up new markets in sectors like construction and healthcare.

9.2 Polyfunctional Robots of the Future

Robots that can fix themselves, comprehend human emotions, and function independently in challenging situations will become commonplace in the field of robotics. To reach previously unheard-of levels of usefulness and adaptability, these next-generation polyfunctional robots will combine state-of-the-art artificial intelligence, cognitive computing, and sophisticated materials.

Autonomous Robots

The idea of self-healing robots is a big step forward for durability and self-maintenance. Self-healing robots, which draw inspiration from biological systems, are able to identify damage and fix it on their own without assistance from humans. This feature is especially helpful in places like space, undersea exploration, or isolated catastrophe areas where routine maintenance is difficult.

Self-healing robots' salient characteristics include:

Self-Detection of Damage:

Real-time detection of cracks, tears, or malfunctions is possible because of sensors included into the robot's framework. The robot's central processing unit receives input from these sensors and starts the repair procedure.

The self-repairing mechanisms are:

These robots can fix small damage by initiating chemical reactions or physical processes using cutting-edge materials like self-healing polymers. For instance, heat-activated polymers that flow and solidify to fill the gap can be used to seal a puncture in a robotic limb.

One of the redundancy systems is

Next-generation robots may feature redundant systems that enable them to continue operating even in the event of partial damage, in situations where self-repair is not practical. Their dependability and resilience in mission-critical tasks are improved as a result.

For extended missions and dangerous situations, self-healing robots will be indispensable since they will minimize downtime, maintenance expenses, and the need for human involvement.

Emotionally Intelligent Autonomous Robots

The creation of emotional intelligence (EI) Robots are another revolutionary concept in polyfunctional robotics. These machines are able to recognize human emotions, react sympathetically, and modify their actions accordingly. Advanced artificial intelligence (AI), natural language processing, and computer vision are used to create emotional intelligence in robotics.

Emotionally intelligent robot applications include:

Medication and Senior Care:

EI-enabled robots can be companions, keep an eye on patients' health, and help with everyday duties. They are able to identify symptoms of perplexity, loneliness, or distress and react appropriately by alerting caretakers or taking necessary action.

Service to Customers:

By comprehending requirements, preferences, and moods, emotionally intelligent robots can improve client experiences in the retail, hotel, and service sectors. They are able to have meaningful conversations, handle conflict amicably, and offer tailored support.

Training and Education:
By adapting their teaching strategies according to student involvement and comprehension levels, robots with emotional intelligence (EI) can facilitate personalized learning. They are able to accommodate various learning methods, identify dissatisfaction, and provide encouragement.

Collaboration at Work:
Emotionally intelligent robots can be productive team players in cooperative work settings by comprehending group dynamics, providing assistance, and settling disputes through flexible communication.

Emotional intelligence will help bridge the gap between technology and human-centric services by making polyfunctional robots more suitable for jobs requiring empathy, social interaction, and adaptive communication.

9.3 Creating Tomorrow's Workforce

The workforce will undergo a fundamental transformation due to the emergence of polyfunctional robots, which will introduce new industries, change current positions, and need the acquisition of new skill sets. Policymakers, businesses, and educational institutions must work together to prepare for these changes.

Getting Ready for Industries Driven by Robots

For workers and organizations to prosper in an economy driven by robotics, proactive adaptation is required. Important tactics for workforce readiness include:

The process of reskilling and upskilling
To operate, maintain, and cooperate with robots, workers must learn new skills. It will be crucial to implement training programs in AI, data analysis, robotics programming, and machine learning.

Emphasis on Soft Skills:

Human roles will prioritize creativity, critical thinking, problem-solving, and emotional intelligence while robots tackle monotonous and technical duties.

Systems of Education That Are Adaptable:

Robotics, coding, and automation principles must be incorporated into curricula at schools and colleges. Programs for lifelong learning must be available to guarantee that employees remain competitive in a labor market that is changing quickly.

Cooperation Between Industry and Academics

Strong collaborations between academic institutions and business executives are necessary for the successful transition to robotics-driven sectors. These partnerships can:

Promote Innovation and Research:

Universities and tech firms can collaborate to create innovative robotic applications and technologies.

Students can gain practical experience in robotics-related professions through internship programs, apprenticeships, and cooperative education projects.

Matching Skills to Industry Requirements:
Educational institutions can guarantee that their curricula meet industry expectations by collaborating closely with businesses, which will close skill gaps and increase employability.

Emerging technologies, ground-breaking inventions, and a flexible workforce are shaping the future of polyfunctional robotics. Robots will open up new opportunities in a variety of industries as they grow more intelligent, robust, and emotionally intelligent. Society can fully utilize robotics to build a more productive, flexible, and cooperative future by adopting these trends.

CHAPTER 10

Using Multipurpose Robots in Your Company

Polyfunctional robots are becoming a more and more useful instrument for increasing productivity, cutting expenses, and raising the standard of services provided by businesses. However, careful preparation, deliberate execution, and constant improvement are necessary for a successful deployment. A thorough method for incorporating polyfunctional robots into your company is described in this chapter, with an emphasis on planning and assessment, deployment tactics, monitoring, and ongoing development.

10.1 Planning and Evaluation

Businesses must perform a comprehensive assessment of their operational needs and capabilities prior to implementing polyfunctional robots. This phase is essential for guaranteeing that automation will actually benefit the company and be in line with long-term goals. The evaluation procedure ought to be methodical and entail giving the duties, expenses, and possible results considerable thought.

Determining Which Tasks Are Automatable

Choosing which business operations are most appropriate for automation by polyfunctional robots is the first stage in the planning phase. It is crucial to concentrate on areas where robots can be most useful because not all tasks are suitable for robotic automation.

Important standards for determining tasks consist of:

Predictability and Repetitiveness:
Automation is best suited for tasks that are predictable, repeatable, and based on distinct patterns. Robots can expedite a variety of jobs, such as inventory management, production quality control, and industrial assembly lines. These jobs frequently call for a high level of accuracy, which robots are better at than people.

Hazardous Conditions and Safety:

When it comes to performing hazardous or dangerous activities that could endanger human workers, robots are ideal. This involves functioning in harsh conditions, like deep-sea research or space missions, or working with hazardous chemicals or large gear. By using robots to perform such activities, worker safety is guaranteed and the chance of harm is reduced.

Decision Making Based on Data:

Large datasets must be processed and analyzed for some commercial operations, like supply chain optimization, market research, and customer behavior analysis. By rapidly evaluating data and providing actionable insights, polyfunctional robots with AI and machine learning skills can greatly improve decision-making.

Low-Complexity, High-Volume Tasks:

Robots can perform large quantities of basic jobs like packaging, sorting, and customer service in industries like retail, logistics, and customer service, freeing up human employees to concentrate on more complex duties like strategic planning or problem-solving.

Businesses can guarantee that robots are placed where they will have the greatest impact by determining which jobs are appropriate for automation, which will eventually increase production and efficiency.

Feasibility and Cost Analysis

It is crucial to carry out a cost and feasibility analysis to assess the financial feasibility of deploying polyfunctional robots after the tasks that can be automated have been determined. Both the initial and continuing expenses as well as the expected long-term benefits should be taken into account in this analysis.

Important things to think about are:

The initial investment costs are as follows:
The implementation of polyfunctional robots necessitates a substantial initial outlay for software, hardware, and system integration. Buying the robots, training employees, and modifying them to fit certain corporate requirements might all be included in the original cost. Evaluating

whether the company can manage this investment and whether the possible return on investment (ROI) warrants the cost is crucial.

Costs of Operation and Maintenance:

Businesses need to budget for recurring running expenses like electricity, software upgrades, and robot maintenance in addition to the original investment. Robots must need routine maintenance and updates in order to function at their best throughout the duration of their lives. Companies need to assess if they can afford these recurring expenses or if outside partners or vendors are required.

The anticipated return on investment, or ROI, is as follows:

One of the most important aspects of the feasibility study is calculating ROI. This entails projecting the cost savings and increased productivity that automation will bring to the company. ROI can be influenced by a number of aspects, including increased speed, lower labor costs, and higher accuracy. Long-term advantages including scalability, expanded capacity, and enhanced customer satisfaction must also be taken into account by businesses.

Flexibility and Scalability: Polyfunctional robots ought to be able to expand with the company. The cost analysis should look at how readily the robot system may be expanded or modified to meet additional duties, growing demand, or changing needs. As their operations develop, companies should also think about whether the robots can be enhanced or combined with other technologies.

Businesses can choose the best implementation method and decide whether polyfunctional robots are a good investment by carrying out a thorough cost and feasibility analysis.

10.2 Strategies for Deployment

Following the completion of the planning and assessment stages, companies can proceed with the deployment of polyfunctional robots. To guarantee that the robots achieve the desired outcomes and blend in seamlessly with current activities, the deployment procedure should be planned and executed gradually.

Testing with Minimal-Scale Executions

Pilot projects are the most efficient approach to deploy polyfunctional robots. Businesses may test the robots in real-world settings, spot possible problems, and improve the system before expanding by starting with small-scale deployments.

Piloting has the following advantages:

Risk Reduction:
Businesses can reduce risks and prevent major disruptions to regular operations by starting small. If issues come up during the pilot stage, they can be fixed before full deployment, avoiding expensive errors or downtime.

Compatibility and Integration Testing:
Robot integration with current systems, including customer relationship management (CRM) tools, enterprise resource planning (ERP) software, and other corporate applications, can be tested during the pilot phase. Maximizing efficiency requires that robots and these systems be compatible and communicate easily.

Performance Measurement:

Businesses should keep a careful eye on robot performance throughout the pilot phase, tracking important parameters including speed, accuracy, uptime, and operational efficiency. Prior to scaling up, gathering this data will assist in improving the system and identifying areas that require development.

Input and Support from Employees:

Pilots provide an opportunity to get input from workers who will coexist with the robots. Acceptance and integration can be enhanced by addressing concerns, providing training and support, and comprehending how staff members engage with the robots. Getting support from employees is crucial to the success of automation initiatives.

Businesses can obtain important insights into the efficacy and performance of polyfunctional robots by conducting small-scale pilots, which will facilitate a more seamless transition to full deployment.

Expanding for Optimal Performance

Businesses can proceed with scaling up the implementation after the pilot phase is successfully finished and modifications have been made. To guarantee that the robots can manage growing workloads and provide optimal efficiency, scaling should be done gradually.

Important factors to take into account when scaling up are:

Capacity and Infrastructure:
Businesses need to make sure their infrastructure can handle the growth as the number of robots rises. This could entail extending physical space to make room for more machines, improving systems, and boosting bandwidth for communication between robots and control centers.

Automation of Other Tasks:
Businesses can think about extending the usage of robots to other jobs or areas of operation once they have demonstrated success in their original responsibilities. Because polyfunctional robots are so adaptable, scaling up

may entail adding additional duties or expanding the robots' reach to include more departments or places.

The process of streamlining workflow:
Workflow optimization is crucial as robots are scaled up to guarantee optimal efficiency. This entails figuring out how humans and robots interact, controlling robot schedules, and making sure that jobs are finished in the most effective manner.

Ongoing Observation:
Continuous monitoring becomes more crucial as companies expand their robot operations. Maintaining high levels of productivity and efficiency requires making sure that robots are operating at their best and adapting to changes in workload or environmental circumstances.

Effective scaling up necessitates meticulous preparation and close attention to detail in order to prevent system overload and guarantee that robots continue to function properly as they take on greater responsibility.

10.3 Observation and Ongoing Enhancement

The deployment of polyfunctional robots is not the end of their implementation. Businesses must use strong monitoring and continuous improvement procedures to guarantee long-term performance. To maintain optimal system performance, this entails monitoring performance, assessing results, and making continuous modifications.

Benchmarks and Performance Metrics

In order to monitor the efficacy of polyfunctional robots, companies need to set important performance metrics and benchmarks. These measures must be in line with corporate goals and concentrate on areas where robots can have the biggest influence.

Among the important performance indicators are:

Completion Time of Task:
Businesses can measure the speed gains from automation by comparing the time it takes robots to accomplish jobs to that of human labor or earlier processes.

Quality Control and Accuracy:

Robots are frequently used to increase consistency and quality. Robots are guaranteed to achieve the required quality standards when the accuracy of their outputs, such as the quantity of errors or faults in production, is monitored.

Uptime of Operations:

Robot uptime monitoring is essential for evaluating system dependability. Any downtime must be minimized and promptly fixed because robots should be accessible and functional for the majority of the day.

Cost Savings:

Cost saving is one of automation's main advantages. Businesses can assess the robots' overall value and return on investment by monitoring labor cost savings, maintenance expenses, and other financial indicators.

Businesses may evaluate the effectiveness of their robotic installation and pinpoint areas for more optimization by tracking these crucial data.

Modernizing Robots for Changing Requirements

Robots with many functions must be able to adjust to the shifting demands of the company. Robots may need to be updated with new hardware, software, or capabilities as the industry develops in order to remain competitive.

Upgrade considerations include:

AI advancements and software updates:
Software is what allows robots to do tasks and make decisions. To guarantee that robots can manage changing jobs and environmental changes, regular software updates including enhancements to AI models are required.

Changes to the Hardware:
Robots may need extra sensors or hardware changes as business requirements change in order to carry out new jobs or adjust to new situations. To travel new areas or do more difficult duties, for example, robots could require improved visual systems.

Getting Ready for New Tasks:

Robots must be retrained to perform new activities or workflows, just as employees must be trained for new duties. Reprogramming or modifying machine learning models to take into consideration fresh data may be necessary for this.

As the organization's demands change, continuous improvement guarantees that robots will continue to provide value and be in line with corporate objectives.

Business use of polyfunctional robots necessitates careful preparation, calculated deployment, and continuous observation. Businesses can successfully incorporate automation into their operations by determining appropriate activities, carrying out feasibility studies, and testing robots before expanding. Businesses can make sure that their robotic workforce stays productive, flexible, and equipped to handle new difficulties by implementing system upgrades, performance tracking, and ongoing monitoring.

ABOUT THE AUTHOR

 Author and thought leader in the IT field Taylor Royce is well known. He has a two-decade career and is an expert at tech trend analysis and forecasting, which enables a wide audience to understand complicated concepts.

Royce's considerable involvement in the IT industry stemmed from his passion with technology, which he developed during his computer science studies. He has extensive knowledge of the industry because of his experience in both software development and strategic consulting.

Known for his research and lucidity, he has written multiple best-selling books and contributed to esteemed tech periodicals. Translations of Royce's books throughout the world demonstrate his impact.

Royce is a well-known authority on emerging technologies and their effects on society, frequently requested as a speaker at international conferences and as a guest on tech podcasts. He promotes the development of ethical technology, emphasizing problems like data privacy and the digital divide.

In addition, with a focus on sustainable industry growth, Royce mentors upcoming tech experts and supports IT education projects. Taylor Royce is well known for his ability to combine analytical thinking with technical know-how. He sees a time when technology will ethically benefit humanity.

www.ingramcontent.com/pod-product-compliance
Lightning Source LLC
Chambersburg PA
CBHW071032240526
45469CB00006BD/2179